Daily *warm-ups*

MATH WORD PROBLEMS

Josh Brackett

The classroom teacher may reproduce materials in this book for classroom use only.
The reproduction of any part for an entire school or school system is strictly prohibited.
No part of this publication may be transmitted, stored, or recorded in any form
without written permission from the publisher.

1   2   3   4   5   6   7   8   9   10
ISBN 0-8251-5066-3
Copyright © 2004
Walch Publishing
P.O. Box 658 • Portland, Maine 04104-0658
walch.com
Printed in the United States of America

Table of Contents

To the Teacher iv

Customary Measurement 1

Metric Measurement 33

Time Measurement 52

Money 62

Counting 111

Probability 173

Answer Key 181

Daily Warm-Ups: Math Word Problems

The Daily Warm-Ups series is a wonderful way to turn extra classroom minutes into valuable learning time. The 180 quick activities—one for each day of the school year—review, practice, and teach math word problems. These daily activities may be used at the very beginning of class to get students into learning mode, near the end of class to make good educational use of that transitional time, in the middle of class to shift gears between lessons—or whenever else you have minutes that now go unused. In addition to providing students with fascinating math activities, they are a natural path to other classroom activities involving critical thinking.

Daily Warm-Ups are easy-to-use reproducibles—simply photocopy the day's activity and distribute it. Or make a transparency of the activity and project it on the board. You may want to use the activities for extra-credit points or as a check on problem-solving skills.

However you choose to use them, *Daily Warm-Ups* are a convenient and useful supplement to your regular lesson plans. Make every minute of your class time count!

Customary Measurement

A helium balloon was let loose from San Francisco, California, which is at sea level. It rose to 10,000 ft and was carried eastward by the winds. Then it lost 4,720 ft of altitude and came down in Denver, Colorado. How high is Denver?

Customary Measurement

A helium balloon rose to 500 ft and was blown sideways for about a mile. Then it went up 350 ft, came down 100 ft, and went up 400 ft. What is the highest altitude the balloon reached?

Customary Measurement

Clint wants to cook something in the oven at 350°F. The room temperature in the kitchen is 70°F. How much does the oven have to heat up before it is ready?

Customary Measurement

Melosa had 2 lb of raisins. She ate 4 oz of them. What fraction of the raisins did she eat?

Customary Measurement

The temperature today in Ellsworth, Maine, is −14°F. In Laredo, Texas, it is 64°F. What is the difference in temperature between the two cities?

Customary Measurement

The paper size commonly used in the United States for business letters, documents, and books is $8\frac{1}{2}$ in by 11 in. What is the area of a sheet of paper of that size?

Customary Measurement

An airplane took off from Cyprus, which is at sea level, climbed to 20,000 ft. It landed at Addis Ababa, which is located in the mountains of Ethiopia at an altitude of 7,900 ft. How much altitude did the plane lose when it landed?

Customary Measurement

By tradition, horse races are measured in furlongs. A furlong is $\frac{1}{8}$ of a mile. How long is a 5-furlong race in miles?

Customary Measurement

Debra bought wall-to-wall carpeting for a 12-ft by 20-ft room. The carpet cost $11.50 per square yard. How much did she have to pay?

Customary Measurement

A rectangle is 4 ft by 9 ft. What is the perimeter of a square that has the same area as the rectangle?

Customary Measurement

In the folktale of Tom Thumb, Tom had a pair of 7-league boots. When he ran wearing the boots, his stride was 7 leagues long. (A stride is the distance from where one foot hits the ground to where the same foot hits the ground again. A league is equal to 3 miles.) How many miles could Tom run in 50 strides?

Customary Measurement

Juanita owns a meat market. She started the day with 60 lb of hamburger. She sold 42 lb. She took home 2 lb for her family. How many pounds did she have left in the store?

Customary Measurement

Paper burns at 451°F. If there is paper in a room where the temperature is 70°F, how much does the temperature have to rise for the paper to catch fire?

Customary Measurement

The average temperature at the South Pole ranges from about –6°F in the summer to –109°F in the winter. How much warmer is the summer than the winter?

Customary Measurement

A penny is 0.05512 in thick. How tall would a stack of 100,000 pennies be?

Customary Measurement

A room has four windows. Each window is 44 in by 60 in. What is the total area of all the windows in the room?

Customary Measurement

At sea level, sound travels at 1,088 ft/sec. If you see lightning strike in the distance and 5 seconds later you hear thunder, approximately how far away is the place where the lightning struck?

Customary Measurement

By tradition, the height of a horse is measured in hands and inches. A hand is 4 in, which is roughly the width of a grown man's hand. Horses are measured from the ground to the highest point of the withers, the ridge between the shoulder bones at the base of the neck. If a horse is 15 hands 2 in tall, how tall is the horse in feet and inches?

Customary Measurement

Dudley is driving from Mateo's house to Mora's house. When he has gone 50 miles beyond halfway, he will be 70 miles from Mora's house. How far is it from Mateo's house to Mora's house?

Customary Measurement

Enrico bought some farmland for $68 per acre. (An acre is 43,560 square feet.) The land he bought is 2,541 ft by 120 ft. How much did he have to pay?

Customary Measurement

Alex is 5 ft $2\frac{1}{2}$ in tall. Julia is 4 ft $7\frac{1}{4}$ in. How much does Julia have to grow to be as tall as Alex is now?

Customary Measurement

Hayley wants to cover the floor of her room with tiles that are 1 ft square. The floor is 12 ft by 20 ft, including a closet that is 3 ft by 3 ft. The floor of the closet will not be covered with tile. How many squares of tile does she need?

Customary Measurement

If a rectangular room is 12 ft wide and its perimeter is 54 ft, how long is it?

Customary Measurement

There are 260,000,000 people in America. If the average American throws away 4.5 lb of trash per day, how much trash does America throw away in a year?

Customary Measurement

On her second birthday, Angela was 2 ft 11 in tall. Today is her 18th birthday, and she is 5 ft 10 in tall. How many inches per year did she grow, on the average, between the two birthdays?

Customary Measurement

Ramon sailed his boat into a harbor where he had never been before. His boat draws 3 ft, which means that the lowest part of the boat is 3 ft below the surface of the water. He had an old chart that said that the harbor was 3 fathoms deep. (A fathom is 6 ft.) If the old chart is still accurate, how deep was the water from the bottom of the boat to the bottom of the harbor?

Daily Warm-Ups: Math Word Problems

Customary Measurement

The equator is a circle that goes around Earth halfway between the North and South poles. The equator is divided into 360° of longitude. Each degree is divided into 60 minutes of arc. Each minute of arc is one nautical mile. (Nautical miles are used all over the planet for air and sea navigation. A nautical mile is not the same as the mile we use on land. One nautical mile is 1.1508 miles.) How many nautical miles long is the equator?

Customary Measurement

Typesetters measure lines of type in picas. A pica is about 4 mm or $\frac{1}{6}$ of an inch. A story that you wrote has 1,581 characters (letters, punctuation marks, and spaces between words) in it. It is going to be printed in a book in type that averages 2.45 characters per pica. Each line of type in the book will be 30 picas long. There will be 28 lines of type on each page. How many pages of the book will it take to print your story?

Customary Measurement

Tuan needed a piece of plywood 39 in by 23 in. He cut it out of a piece that was 4 ft by 8 ft. What is the area in square inches of the piece that was left?

Customary Measurement

Benson bought a piece of cloth 1 yd wide by $5\frac{1}{2}$ yd long. He cut 3 in off the entire length of the long side. What is the area of the piece of cloth he had left?

Customary Measurement

Vanessa needs to pave an 8-ft by 50-ft driveway with 2 in of concrete. How many cubic yards of concrete does she need?

Customary Measurement

An airplane flew from Asmara, Ethiopia, where the latitude is 15° North, to Tananarive, Madagascar, where the latitude is 19° South. What is the difference in latitude between the two cities?

Daily Warm-Ups: Math Word Problems

Metric Measurement

If the pages of a 256-page book are 1.365 cm thick, how thick is one page in millimeters?

Metric Measurement

A group of mountaineers climbed to the summit of Mount Everest, which is 8,850 m above sea level. How far did they have to climb down to return to the base camp at 5,300 m?

Metric Measurement

Brian opened a 538-g can of chicken soup and ate half of it. How much soup did he eat?

Metric Measurement

Heike Drechsler of Germany won the gold medal in the women's long jump at the 2000 Olympics in Sydney, Australia. She jumped 6.99 m. Four years earlier, Chioma Ajunwa of Nigeria won by jumping 7.12 m in the same event. How much farther in centimeters did Ajunwa jump than Drechsler?

Metric Measurement

Sarah dove off the 10-m platform into the pool. She went 2 m below the surface before she started to come up. How far from the platform was she then?

Metric Measurement

A missile was fired from a submarine floating 200 m below the surface of the ocean. It rose to a height of 500 m above sea level before coming down. How high above the submarine did the missile go?

Metric Measurement

One day last winter, the temperature was 10°C in the early afternoon. By morning, it had gone down to –7°C. By how many degrees did the temperature drop?

Metric Measurement

A commercial jet flies from St. Louis, Missouri, to Buenos Aires, Argentina, a distance of 4,707 nautical miles. If the plane flies at 210 knots, how long will the trip take? (A knot is 1 nautical mile per hour.)

Metric Measurement

The owner of a dairy farm has to fence off a rectangular area 143 m by 63 m to pasture cows. How long will the fence be?

Metric Measurement

The paper size commonly used in Europe for business letters, documents, and books is 210 mm by 297 mm. What is the area of a sheet of paper of that size?

Metric Measurement

Delilah is making a collage. She is going to paste 4 strips of 35-mm film negatives side-by-side onto a piece of paper. How many centimeters wide will the 4 strips be?

Metric Measurement

In May 2001, the Ukraine produced 1.468 Gm^3 of natural gas. In May 2002, the Ukraine produced 1.534 Gm^3. How much more natural gas did they produce in May 2002 than in May 2001?

Metric Measurement

The nurse took Linda's temperature. It was 37.5°C, which was 17.5°C warmer than the temperature in the room. What was the temperature in the room?

Metric Measurement

The speed limit on major highways in Germany is 130 km/hr. How fast is that in miles per hour? (1 km = 0.6 miles)

Metric Measurement

The Texas Instruments Company makes a computer chip for cellular telephones that is 90 nanometers (nm) long. How many of these chips laid end-to-end would it take to reach from one end of a meter stick to the other?

Metric Measurement

The white keys on Lesley's piano are 2 cm wide. The black keys are 1 cm wide. If an octave has 7 white keys and 5 black keys, how wide is an octave? (Imagine that the keys are side by side.)

Metric Measurement

Omar, Ben, and Darnell all live along a straight road in the desert. Ben called Omar and said, "My car won't start, and I have to get to Darnell's house. Can you give me a ride?" Omar said, "Yes." Omar drove 18 km to Ben's house. Then he drove 39 km back past his own house to Darnell's house and dropped Ben off. How far is it from Omar's house to Darnell's?

Metric Measurement

The land area of the United States is 9,158,960 km². How many hectares is that? (A hectare is 10,000 m².)

Metric Measurement

Wanda bought some farmland for $11.50 per hectare. (A hectare is 10,000 m^2.) The land she bought is 3,800 m by 60 m. How much did she have to pay?

Time Measurement

Amy, Barbara, Charlie, David, and Everett all play the drums. Amy beats her drum once every second; Barbara, every other second; Charlie, every third second; David, every fourth second; and Everett, every fifth second. If they all start by beating their drums together at the first second, how long will it take before they all beat their drums together again?

Time Measurement

The Greek philosopher and mathematician Pythagoras lived from 569 B.C.E. to 475 B.C.E. How long did he live?

Time Measurement

The school bus comes to pick up Bailey at 7:30 A.M. It takes him 10 min to get up and get dressed, 20 min to have breakfast, and 15 min to walk to the bus stop. What time does he have to get up?

Time Measurement

The television footage showed the rocket launch from T – 1 min 45 sec to T + 3 min 30 sec. How long did the footage last?

Time Measurement

Darcy's band recorded a song that lasted 3 min 23 sec. They decided it was too short, so they added another 2 min 49 sec to it. How long is the new version?

Time Measurement

Darcy's band recorded a song that lasted 7 min 13 sec. They decided it was too long, so they cut 1 min 22 sec out of it. How long is the cut version?

Time Measurement

Pedro shot a roll of 36 pictures. For each of them, the film was exposed for $\frac{1}{125}$ of a second. What was the total exposure time of the entire roll?

Time Measurement

To make the dinner Lian wants to cook, the oven has to preheat for 10 min. The dinner has to cook for 45 min. It will take him 15 min to prepare the dish before putting it in the oven. The directions say to let it stand for 10 min before serving. Dinner has to be ready at 6:30 P.M. What time does Lian have to start preparing dinner?

Time Measurement

Darcy's band's new CD has 13 songs on it. The playing times of the songs are 5:23, 5:48, 5:51, 5:27, 5:52, 5:56, 4:33, 3:45, 5:25, 5:30, 5:27, 6:14, and 5:59. (5:23 means 5 min 23 sec.) What is the total playing time of all the songs on the CD?

Time Measurement

Two elevators going down leave the 10th floor at the same time. One elevator is faster than the other. The faster elevator takes 45 sec between floors. The slower elevator takes 1 min between floors. There are people on every floor waiting to go down. The first elevator that gets to their floor must stop to pick them up. It takes 3 min for either elevator to stop and pick up passengers. Which elevator will reach the first floor first?

Money

Mr. Morgan wanted Oswald to do some work for him. Oswald made the following agreement with him. "I'll work for you for 10 days," Oswald said. "You only have to pay me $1 for the first day, $2 for the second day, $4 for the third day, and so on." How much did Mr. Morgan have to pay Oswald for the 10 days?

Money

Lisa bought a computer. The list price was $699. Because it was an old model, it was on sale at 20% off. She had to pay a 5% sales tax. How much did she have to pay for the computer?

Money

Ryan owns a clothing store. He bought two jackets in the same style. The list price was $80 each. He bought them at 45% off list price. He sold one for full list price right away, but the other one didn't sell for a long time. He marked it down and ended up selling it for $5 more than he paid for it. How much profit did he make on the two jackets combined?

Money

According to his July 10 statement, Kobe owes $300 on his credit card account. The credit card company charges 18% per year ($1\frac{1}{2}$% per month) for a finance charge. He sends in a $50 payment that arrives just before his August 10 statement is printed. What will the balance on his August 10 statement be? As a percentage, what part of his $50 payment went to pay interest rather than to reduce Kobe's balance?

Money

According to her October 5 statement, Felicia owes $500 on her credit card account. She sends in the minimum payment, which is $10. It arrives just before the November 5 statement is printed. The credit card company charges 18% per year ($1\frac{1}{2}$% per month) for a finance charge. What will the balance be on her November 5 statement? As a percentage, what part of the $10 payment went to pay interest rather than reduce Felicia's balance?

Money

William lives in Massachusetts, where there is a 5% sales tax on everything sold in stores except food and clothing. He went to a store and bought a shirt for $29.99, a CD for $18.99, and a pen for $2.00. How much did he have to pay?

Money

Marilyn filled her car up with gasoline. She bought 13.8 gal at $1.75/gal. How much did she have to pay?

Money

Marilyn filled her car up with gasoline. She bought 13.8 gal. Her trip mileage indicator shows that she had gone 290.1 miles since the last time she filled the tank. How many miles per gallon did her car average?

Money

One bottle of a certain soft drink costs $1.25. A case of 24 costs $25. As a percentage, how much do you save by buying by the case?

Money

A pint of ice cream costs $2.59. A 5-gal container costs $77. As a percentage, how much do you save by buying a 5-gal container?

Money

Erik was arrested for driving without a license. He was fined $500 and had to pay the attorney who represented him in court $750. He had $273.15 in his savings account. He had to borrow the rest. How much did he have to borrow?

Money

Abby likes to read a certain magazine every month. The newsstand price is $1.75 per issue. A year's subscription is $12.99. As a percentage, how much will she save by buying a year's subscription?

Money

Ray rented a car for 5 days. The charge was $55 per day plus a 10% surcharge for not returning the car to the same place where he originally got it. How much did he have to pay to rent the car?

Money

Grace bought a valuable sports card at an online auction for $70 plus a shipping charge of $9.00. She later sold it for $105. As a percentage, how much profit did Grace make?

Money

Mr. and Mrs. Rodriguez bought their house in the spring of 2000 for $215,000. They sold it 3 years later for $280,000. As a percentage, how much did the price of the house increase per year?

Money

Andrew bought a famous basketball player's autograph at an on-line auction for $130 plus a shipping charge of $9.00. Later, the basketball player was convicted of a crime. Andrew sold the autograph for $22. As a percentage, how much did he lose?

Money

Darcy's band played in a concert. Her agreement with the theater was that the band would get 50% of gross ticket sales with $300 guaranteed. (Gross sales means total sales before any expenses are deducted.) The theater sold 98 tickets at $12 each. How much was the band paid?

Money

Darcy's band played in a concert. Her agreement with the theater was that the band would get 50% of gross ticket sales with $300 guaranteed. (Gross sales means total sales before any expenses are deducted.) A very well-known band was playing nearby. Only 18 tickets were sold at $12 each. How much was the band paid?

Money

Martin put $123 in a savings account and left it there for 3 years. It earned interest at 2% per year compounded annually. What was the balance after 3 years?

Money

Akira bought a skateboard online for $69.99. The shipping charges were $17.98. What was the total amount he had to pay?

Money

Bryce had some money of his own. His mother gave him $15. He went to a store and spent $80. How much of his own money did he spend?

Money

Caleb has $100. Using his credit card, he buys a shirt for $48. He gets $25 for his birthday. Using his credit card again, he buys a pair of pants for $30. He returns the shirt to the store. The credit card company charges him $1.42 interest. How much money will he have after he pays the bill?

Money

Ling has 81 dimes and 29 quarters. How much money does she have?

Money

Doug went shopping and spent $1 on Monday. On Tuesday, he spent 3 times as much as on Monday. On Wednesday, he spent 3 times as much as on Tuesday, and so on through the week. How much did Doug spend on Saturday?

Money

Mya had $15. She gave one third of it to her mother, 40% of it to her brother, and $4 to her sister. Who got the most? Who got the least?

Money

Isabelle had a balance in her checking account of $768.44. She made a deposit of $79.40 and wrote a check for $69.67. What was her new balance?

Money

Kevin had a balance in his checking account of $986.60. He made two deposits: one for $62.59 and one for $82.68. What was his new balance?

Money

Ahmed had a balance in his checking account of $171.50. He made a deposit of $26.79 and wrote a check for $86.29. What was his new balance?

Money

Taylor had a balance in her checking account of $857.11. She wrote a check for $89.23 and a check for $29.92. What was her new balance?

Money

Emma bought 11 gal of gasoline at $1.69/gal. How much did she have to pay?

Money

Mercedes is saving $2.25 a week from her allowance to buy a CD that costs $15. How many weeks will she have to save before she can buy the CD? How much will she have left over?

Money

Miguel won $10,000,000 in the lottery. He had to pay $3,800,000 in federal income taxes. How much did he have left?

Money

The letter carrier comes to your house and says, "I made a mistake. Remember that envelope I brought you yesterday? Well, it was meant for somebody else, so I have to take it back." In the envelope there was a check for $55. Are you better off or worse off? By how much?

Daily Warm-Ups: Math Word Problems

Money

As a promotion, a store let the first 99 customers on a certain day buy a 3-liter bottle of pure olive oil for only 99 cents each. The bottles cost the store $4.50 each. How much did the promotion cost the store?

Money

Avi is going to Zurich, Switzerland. He is bringing $300 with him to buy gifts for his family. He goes to the bank and gets his money changed into Swiss francs (SFr). The exchange rate is $1 = SFr6.35. How many Swiss francs will he get to spend in Zurich?

Money

Katie had so many quarters she could arrange them in a square with 4 quarters on each side. How much money did she have in dollars?

Money

Dominic has 13 coins. Together they are worth $1.65. One of them is a 50-cent piece. What are the others?
Hint: Make a chart.

Money

Maggie just got back to the United States from her trip to London. She has some British pounds (£) left that she did not spend. She goes to the bank and gets £18.50 changed into U.S. dollars. The exchange rate is £1 = $1.81. How much will she get in U.S. currency?

Money

Paulette is coming home to the United States from Japan. She has some Japanese yen (¥) left that she did not spend. She goes to the bank and gets ¥5,413 changed into U.S. dollars. The exchange rate is ¥1 = $0.0034. How much will she get in U.S. currency?

Daily Warm-Ups: Math Word Problems

100

Money

Paulette is going from Japan to France. She is bringing Japanese yen (¥) with her. She has ¥73,850. She goes to the bank and gets it changed into euros (€), the currency of the European Union. The exchange rate is ¥134.690 = €1. How much money will she get to spend in France?

Money

Johanna has $61 in seven bills of various denominations. One of the bills is a $20 bill. What are the others? *Hint:* Make a chart.

Money

Yukiko has 14 coins. Together they are worth $1.96. One of them is a 50-cent piece. What are the others?
Hint: Make a chart.

Money

Darcy and her band played in a concert. They were paid $1,000. After paying for $100 in expenses, they divided the money equally among the five members of the band. How much did each member get?

Daily Warm-Ups: Math Word Problems

104

Money

Midori has 14 coins. Together they are worth $2.45. One of them is a 50-cent piece. What are the others?
Hint: Make a chart.

Money

Muhammed has $623 in 20 bills of various denominations. One of the bills is a $20 bill. What are the others? *Hint:* Make a chart.

Money

Carlos and three of his friends ordered food from a restaurant and had it delivered. The bill was $24.80, and they gave the delivery person a $4 tip. If they share equally, how much should each person pay?

Money

Kelly has $53.03 in 17 bills and coins of various denominations. One of them is a $20 bill. What are the others? *Hint:* Make a chart.

Money

Julio went shopping and spent $1 on Monday. On Tuesday, he spent 3 times as much as on Monday. On Wednesday, he spent 3 times as much as on Tuesday, and so on through the week until and including Saturday. How much did he spend during the whole week?

Money

Because of inflation, a diamond ring that sold for $200 in 1950 would sell for $1,500 today. Therefore, today's dollar is worth what percent of the value of a 1950 dollar?

Counting

In a survey of 120 sixth graders, 91 said they liked spaghetti, 88 said they liked meatballs, and 75 said they liked both spaghetti and meatballs. How many did not like either spaghetti or meatballs?

Counting

There are 895 students at the King School. There are 33 more girls than there are boys. How many girls are there?

Counting

Geraldo has 13 pennies. He wants to stack them in 3 piles so that each pile always has an odd number of pennies in it. How many ways can he do that?

Counting

Agnes's father is 6 times as old as she is. In 4 years, he will be 4 times as old. How old are they now?

Counting

A candy company wants to sell chocolates in rectangular boxes of 36. A box that holds one row of 36 would be too difficult to handle. What other rectangular arrays of 36 chocolates are possible?

Counting

Gertrude has a fruit store. She likes to display oranges by stacking them up in a pyramid with 1 orange on top supported by 4 oranges underneath it, which are supported by 9 oranges underneath them, and so on. She stacks the oranges up 7 layers high. How many oranges are there in the display?

Counting

In a class of 30 students, 14 are boys. If 2 boys drop out of the class and 2 girls are added, what fraction of the class will be boys?

Counting

In an election, Ms. Duval got 534 votes, Mr. Smith got 507, and Mr. Fennig got 88. What percent of the vote did the winner get?

Counting

In a class of 28 students, 25% are 11 years old. The rest are 12. If four of them have their 12th birthdays, what will the percentage of 11-year-olds be?

Counting

In a box of 36 chocolates, $\frac{1}{3}$ of them are white; the rest are brown. If 1 white chocolate and 3 brown chocolates are eaten, what fraction of the remaining chocolates will be white?

Counting

In 301 times at bat, Rachel got 69 hits. What is her batting average? (Batting averages are traditionally represented by a decimal with three significant figures.)

Counting

Twelve people go to a party. As each person arrives, he or she shakes hands once with each person who is already there. How many handshakes will take place?

Counting

In an orchestra of 50, $\frac{1}{2}$ of the instruments are strings, $\frac{2}{5}$ are winds, and $\frac{1}{5}$ percussion. If five more percussion instruments are added for a special concert, what fraction of the orchestra will the percussion section be?

Counting

You've been asked to design a case to hold 24 cans of soft drinks. To make it easy to handle, the case should not be long and thin. The dimensions of the case should be as short as possible. How many layers of cans would you have? How many cans on each layer? Arranged how?

Counting

A zoo had 28 rabbits. Then 9 of them died, and 14 more were born. How many rabbits were there then?

Counting

Benny ate $\frac{1}{3}$ of a 15-oz bag of cookies. How much did he eat?

Counting

Darrell has 537 ants in his ant farm. Each of them has 6 legs. How many legs is that?

Counting

Eggs are sold wholesale in cases of 30 dozen. How many eggs are there in 20 cases?

Counting

Bernard had 656 sports cards. Arnie gave him 98 more. He gave 56 to Carter. Then he lost some. He had 253 left. How many did he lose?

Counting

Carter has 438 sports cards. He wants to store them in packages of 24. How many packages does he need?

Counting

Eggs are usually sold to consumers by the dozen in cartons that hold a rectangular array of two rows of six eggs in each row.

An egg company wants to sell eggs to restaurants in rectangular cartons that hold 80 eggs. A carton that holds one row of 80 eggs would be too difficult to handle. What other rectangular arrays of 80 eggs are possible?

Counting

Eggs are usually sold to consumers by the dozen in cartons that hold a rectangular array of two rows of six eggs in each row.

An egg company wants to sell eggs to restaurants in rectangular cartons that hold 90 eggs. A carton that holds one row of 90 eggs would be too difficult to handle. What other rectangular arrays of 90 eggs are possible?

Counting

Anthony wants to plant carrots. He made 3 rows of holes with 18 holes in each row. He wants to plant 5 seeds in each hole. How many seeds does he need?

Counting

If you have 13 blocks, you can stack up 12 of them to make a 2-by-2-by-3 rectangular solid with more than one block on every side. But you cannot make a rectangular solid with more than one block on every side using all 13 blocks. If you have 19 blocks, what is the greatest number of blocks you can stack up to make a rectangular solid with more than one block on every side?

Counting

In a dog-training class of 24, $\frac{1}{2}$ the dogs are terriers, $\frac{1}{6}$ are hounds, and the rest are retrievers. How many retrievers are there?

Counting

It's almost Thanksgiving. A farm has 530 turkeys ready to go to market. They are to be transported in containers that hold no more than 25 turkeys each. How many containers will be needed to transport *all* the turkeys?

Counting

Joseph needed 36 tomatoes for a big batch of spaghetti sauce. He got 10 from his own garden and 9 from a friend's garden. One fell on the floor and got stepped on. How many did he have to buy?

Counting

Kendra had 24 worms. She used 3 as bait. She gave away 5 to a friend. Some of them fell into the water when she accidentally knocked the bait can over. She had 10 left. How many fell into the water?

Counting

Holly wants to plant beans. She made 5 rows of holes with 20 holes in each row. She wants to plant 3 seeds in each hole. How many seeds does she need?

Counting

Syd wants to plant radishes. She made 6 rows of holes with 10 holes in each row. She wants to plant 3 seeds in each hole. How many seeds does she need?

Counting

Neva is planning a party for 25 people. She wants to provide 3 cookies per person. She has 2 dozen on hand. How many more cookies does she need?

Counting

On a test with 30 questions, Shannon got 27 right. What was her score as a percentage?

Counting

On a test with 40 questions, Jamie got 34 right. What was his score as a percentage?

Counting

On a test with 53 questions, Bartholomew got 40 right. What was his score as a percentage?

Counting

Peggy has 175 pictures that she wants to put into albums. Each album has 32 pages in it, and she will put 4 pictures on each page. How many albums does she need? How many blank pages will be left?

Counting

Elizabeth is a very good photographer who takes a lot of pictures. She finds that only about 7% of the pictures she takes are good enough to keep. Recently she shot 3 rolls of film with 36 pictures on each roll. About how many of those do you think she will want to keep?

Counting

Sean needed a lot of cabbage for a big stew. He bought 3 heads of cabbage at one store, 7 at another store, and 7 at another store. Two of them spoiled. How many heads of cabbage did Sean have for the stew?

Counting

The U.S. Postal Service reports that during the mid-1980s more than 7,000 letter carriers were attacked by dogs every year. Because of aggressive employee training and public education programs, that statistic has dropped by more than $\frac{2}{3}$. Approximately how many letter carriers are attacked by dogs now?

Counting

A group of fans of a football team has 64 members. They all wear shirts of the same color. When they go to games, they all sit together in their own section of the stadium with 8 rows of 8 seats each, forming a square. Just before today's game, they find that one member can't come. They only have 63 members and 63 shirts, so they can't form a square. But they want to arrange themselves in a rectangle that is as close to a square as possible. How can they do that?

Counting

Arnie had 984 sports cards. He got tired of them. He sold half of them to a dealer and divided those that were left evenly among his 3 best friends. How many did each friend get?

Counting

Josh has 438 sports cards. Each of them measures $3\frac{1}{2}$ by $2\frac{1}{2}$ inches. If he laid them out on the floor next to each other with no gaps and no overlapping, what area would they cover?

Counting

Josh has 438 sports cards. Each of them measures $3\frac{1}{2}$ by $2\frac{1}{2}$ inches. If they were laid end-to-end the long way with no gaps and no overlapping, how far would they reach?

Counting

Tennis balls are often sold by the dozen in cartons that hold a rectangular array of 3 rows of 4 balls in each row, all on one layer. A sports equipment company wants to sell tennis balls to large sports centers in rectangular cartons that hold 80 balls. A carton that holds one row of 80 balls all on one layer would be too difficult to handle. What other rectangular solid arrays of 80 balls on one or more layers are possible?

Counting

If you have 13 blocks, you can stack up 12 of them to make a 2-by-2-by-3 rectangular solid with more than one block on every side. But you cannot make a rectangular solid with more than one block on every side using all 13 blocks. If you have 41 blocks, what is the greatest number of blocks you can stack up to make a rectangular solid with more than one block on every side?

Counting

If you have 8 cubic blocks, you can stack them up to make a 2-by-2-by-2 cube. If you have 200 cubic blocks, what is the greatest number of blocks you can stack up to make a cube—a stack of which the width, length, and height are the same number of blocks?

Counting

If you have 8 cubic blocks, you can stack them up to make a 2-by-2-by-2 cube. If you have 30 cubic blocks, what is the greatest number of blocks you can stack up to make a cube—a stack of which the width, length, and height are the same number of blocks?

Counting

A new car costs $11,600. The dealer gives you the choice of paying cash or financing it. Paying cash means paying the total cost in full. Financing it means paying $2,000 up front and paying 12 monthly payments of $844. How much is the total cost of the car if you finance it?

Counting

How many 3-by-5-in cards do you need to cover a rectangle 2 ft by 5 ft?

Counting

A 320-acre farm produced 44,800 bushels of corn. What was the average production per acre?

Counting

June 7 and June 14, 2004, are consecutive Mondays. Their sum is 7 + 14 = 21. What is the least sum that two consecutive Mondays can have in any month? What is the greatest?

Counting

Keisha is making a birthday card for a friend. She has 40 interesting stickers. She wants to paste as many stickers as she can on a piece of paper in a triangular design: one sticker at the top, two in the second row, three in the third row, and so on. What is the greatest number of stickers she can use out of the 40 she has?

Counting

Nick and Brady have a newspaper delivery route with 30 customers. They deliver newspapers to each customer 6 days a week all year round. How many newspapers do they deliver in a year?

Counting

Adam took a multiple-choice exam with 90 items on it. He got 81 items correct. He got 5 wrong. He left 4 blank. The examiners give 4 points for every correct answer. They take off 1 point for every wrong answer. They take nothing off for answers left blank. What was Adam's score?

Counting

Soldiers march in rows and columns, but every row and every column has to have the same number of soldiers in it. For example, a squad of 12 soldiers could march in a formation of three rows and four columns or a single column of 12. In how many different formations could a squad of 12 soldiers march?

Counting

The game of bowling is played with 10 pins arranged in a triangle with one pin in the first row, two in the second row, three in the third row, and so on. Suppose you wanted to invent a new kind of bowling game to be played with 60 pins or fewer arranged in a triangle. What is the greatest number of pins you could arrange in a triangle?

Counting

Arabella jogged 500 yd the first day. She increased her distance by 100 yd every day until she reached a mile. How many days did she jog? (1 mi = 1,760 yd.)

Counting

In the Rodriguez family, there are 3 children. The sum of their ages is 10. The product of their ages is 30. What are their ages?

Counting

A doctor, a lawyer, and a minister are friends. Their names are Aiko, Basil, and Camilla. Aiko is not a lawyer. Basil is not 35 years old, but one of his other two friends is. Camilla is not a doctor. The lawyer is not 40, but one of the lawyer's friends is. Camilla is 45. Basil is a minister. What is each person's age and occupation?

Counting

A theater has 20 seats in the first row, 21 in the second row, 22 in the third, and so on. There are 18 rows. How many seats are there in the theater?

Counting

My scores on my first four tests in history were 88, 85, 60, and 95. What score do I have to get on the next test to have an average score of 80?

Counting

This year, Moira's age is a multiple of 7. Next year it will be a multiple of 5. Her older sister is now 18. How old is Moira this year?

Counting

This year, Gaspar's age is a multiple of 5. Next year it will be a multiple of 6. Gaspar's older brother is 37. How old is Gaspar this year?

Probability

There are 75 gumballs in a bag. There are 30 red gumballs, 20 green, 15 yellow, 5 blue, and 5 orange. If you reach into the bag without looking and take out one gumball, what is the probability that you will get a green gumball?

Probability

There are 75 gumballs in a bag. There are 30 red gumballs, 20 green, 15 yellow, 5 blue, and 5 orange. If you reach into the bag without looking and take out 2 gumballs, what is the probability that one of them will be green?

Probability

There are 75 gumballs in a bag. There are 30 red gumballs, 20 green, 15 yellow, 5 blue, and 5 orange. If you reach into the bag without looking and take out 2 gumballs, what is the probability that you will get 2 green gumballs?

Probability

There are 75 gumballs in a bag. There are 30 red gumballs, 20 green, 15 yellow, 5 blue, and 5 orange. If you reach into the bag without looking, what is the probability that you will get a gumball that is not green?

Probability

There are 75 gumballs in a bag. There are 30 red gumballs, 20 green, 15 yellow, 5 blue, and 5 orange. If you reach into the bag without looking, what is the probability that you will get a red or a green gumball?

Probability

A standard deck of playing cards has 52 cards. No two cards are alike. Each card has both a value and a suit. There are 4 suits: hearts, diamonds, clubs, and spades. Each suit has 13 cards in it with values of 2, 3, 4, 5, 6, 7, 8, 9, 10, jack, queen, king, and ace. If you choose a card at random, what are the chances that you will choose a heart?

Probability

A standard deck of playing cards has 52 cards. No two cards are alike. Each card has both a value and a suit. There are 4 suits: hearts, diamonds, clubs, and spades. Each suit has 13 cards in it with values of 2, 3, 4, 5, 6, 7, 8, 9, 10, jack, queen, king, and ace. If you choose a card at random, what are the chances that you will choose a king?

Probability

A standard deck of playing cards has 52 cards. No two cards are alike. Each card has both a value and a suit. There are 4 suits: hearts, diamonds, clubs, and spades. Each suit has 13 cards in it with values of 2, 3, 4, 5, 6, 7, 8, 9, 10, jack, queen, king, and ace. If you choose 2 cards at random, what are the chances that you will choose 2 kings?

Answer Key

Customary Measurement

1. 10,000 ft − 4,720 ft = 5,280 ft
2. 500 ft + 350 ft − 100 ft + 400 ft = 1,150 ft
3. 350°F − 70°F = 280°F
4. 2 lb × (16 oz/lb) = 32 oz. 4 oz/32 oz = 1/8
5. 64°F − (−14°F) = 78°F
6. 8.5 in × 11 in = 93.5 in^2
7. 20,000 ft − 7,900 ft = 12,100 ft
8. 5 furlongs × ($\frac{1}{8}$ mile/furlong) = $\frac{5}{8}$ mile
9. 12 ft × 20 ft × (1 yd^2/9 ft^2) × ($11.50/yd^2) = $306.67
10. Area = 4 ft × 9 ft = 36 ft^2 = 6 ft × 6 ft
    Perimeter of square = 4 × 6 ft = 24 ft
11. 50 strides × (7 leagues/stride) × (3 miles/league) = 1,050 miles
12. 60 lb − 42 lb − 2 lb = 16 lb
13. 451°F − 70°F = 381°F
14. −6°F − (−109°F) = 103°F
15. 0.05512 in/penny × 100,000 pennies = 5,512 in
16. 4 × (44 in × 60 in) = 10,560 in^2
17. 5 sec × (1,088 ft/sec) = 5,440 ft, a little more than a mile
18. 15 hands × (4 in/hand) × (1 ft/12 in) + 2 in = 5 ft 2 in
19. 50 miles + 70 miles = halfway = 120 miles. 2 × 120 miles = 240 miles
20. 2,541 ft × 120 ft × (1 acre/43,560 ft^2) × ($68/acre) = $476
21. 5 ft 2$\frac{1}{2}$ in − 4 ft 7$\frac{1}{4}$ in = 62$\frac{1}{2}$ in − 55$\frac{1}{4}$ in = 7$\frac{1}{4}$ in
22. (12 ft × 20 ft) − (3 ft × 3 ft) = 240 ft^2 − 9 ft^2 = 231 ft^2 = 231 tiles
23. (54 ft − 2 × 12 ft)/2 = 15 ft
24. (4.5 lb/day) × (365 day/year) × 260,000,000 = 427,050,000,000 lb/year
25. 5 ft 10 in − 2 ft 11 in = 70 in − 35 in = 35 in. 18 years − 2 years = 16 years. 35 in/16 year = 2.1875 in/year
26. 6 ft/fathom × 3 fathoms = 18 ft. 18 ft − 3 ft = 15 ft
27. 360° × (60 min/°) = 21,600 min = 21,600 nautical miles
28. (1,581 cc)/(2.45 cc/pica) = 645 picas. 645 picas/(30 picas/line) = 21.5 lines. The whole story will fit on one 28-line page.

Daily Warm-Ups: Math Word Problems

Answer Key

29. 4 ft × 8 ft = 48 in × 96 in = 4,608 in². 39 in × 23 in = 897 in². 4,608 in² – 897 in² = 3,711 in²
30. (1 yd × 36 in/yd) × (5½ yd × 36 in/yd) – (5½ yd × 36 in/yd × 3 in) = 7,128 in² – 594 in² = 6,534 in²
31. 8 ft × (12 in/ft) = 96 in. 50 ft × 12 in/ft = 600 in. 96 in × 600 in × 2 in = 115,200 in³.
    1 yd³ = 36 in/yd × 36 in/yd × 36 in/yd = 46,656 in³/yd³). 115,200 in³/(46,656 in³/yd³) = 2.47 yd³
32. 15° – (–19°) = 34°

Metric Measurement

33. 1.365 cm/256 = 0.00533 cm = 0.533 mm
34. 8,850 m – 5,300 m = 3,550 m
35. 538 g × (½) = 269 g
36. 7.12 m – 6.99 m = 0.13 m = 13 cm
37. 10 m + 2 m = 12 m
38. 500 m – (–200 m) = 700 m
39. 10°C – (–7°C) = 17°C
40. 4,707 miles/(210 miles/hr) = 22.41 hr, almost 23 hr
41. (2 × 143 m) + (2 × 63 m) = 412 m
42. 210 mm × 297 mm = 62,370 mm²
43. 4 × 35 mm = 140 mm = 14 cm
44. 1.534 Gm³ – 1.468 Gm³ = 0.066 Gm³
45. 37.5°C – 17.5°C = 20°C
46. 1 km = 0.6 miles. 130 km/hr × (0.6 km/mile) = 78 miles/hr
47. 1 m = 1,000,000,000 nm. 1,000,000,000 nm/90 nm = 11,111,111.111…
48. 7 × 2 cm + 5 × 1 cm = 14 cm + 5 cm = 19 cm
49. 18 km – 39 km = –21 km. It is 21 km from Omar's house to Darnell's.
50. 9,158,960 km² × (100 ha/km²) = 915,896,000 ha
51. 3,800 m × 60 m × (1 ha/10,000 m2) × ($11.50/ha) = $262.20

Time Measurement

52. 59 seconds. They will all beat together on the 60th beat. (60 is the least common multiple of 1, 2, 3, 4, and 5.)
53. –569 – (–475) = 94 years
54. 7:30 A.M. – 10 min – 20 min – 15 min = 7:30 A.M. – 45 min = 6:45 A.M.
55. 3 min 30 sec – (–1 min 45 sec) = 4 min 75 sec = 5 min 15 sec

Daily Warm-Ups: Math Word Problems

Answer Key

56. 3 min 23 sec + 2 min 49 sec = 5 min 72 sec
    = 6 min 12 sec
57. 7 min 13 sec − 1 min 22 sec = 6 min 73 sec −
    1 min 22 sec = 5 min 51 sec
58. 36 pix × [(1 sec/125)/pix] = 36 sec/125 or
    0.288 sec
59. Lian can prepare dinner while the oven is preheating. 6:30 − 15 min − 45 min − 10 min = 6:30 − 70 min = 5:20 P.M.
60. 5:23 + 5:48 + 5:51 + 5:27 + 5:52 + 5:56 + 4:33 + 3:45 + 5:25 + 5:30 + 5:27 + 6:14 + 5:59 = 63:490 = 71:10
61. The fast elevator has to stop and pick up passengers at the ninth, seventh, fifth, and third floors. The slow elevator stops at the eighth, sixth, fourth, and second floors. The fast elevator makes the trip in 18 min 45 sec. The slow elevator takes 21 min.

Money

62.
Day	Pay (in $)
1	1
2	2
3	4
4	8
5	16
6	32
7	64
8	128
9	256
10	512
Total	$1,023

63. ($699 − 0.20 × $699) × 1.05 = $587.16
64. $80 + [$80 × (1 − 0.45) + $5] − 2 × $80 × (1 − 0.45) = $41
65. $300 × 1.015 − $50 = $254.50. $4.50 or 9% went to pay interest.
66. $500 × 1.015 − $10 = $497.50. $7.50 or 75% went to pay interest.
67. ($18.99 + $2.00) × 1.05 + $29.99 = $52.03
68. 13.8 gal × $1.75/gal = $24.15

Daily Warm-Ups: Math Word Problems

Answer Key

69. 290.1 miles/13.8 gal = 21.02 miles/gal
70. (24 × $1.25 − $25)/(24 × $1.25) = $16\frac{2}{3}$%
71. (5 gal × 2 pt/qt × 4 qt/gal × $2.59/pt − $77)/(5 gal × 2 pt/qt × 4 qt/gal × $2.59/pt) = 26%
72. $500 + $750 − $273.15 = $976.85
73. (12 × $1.75 − $12.99)/(12 × $1.75) = 38%
74. 5 × $55 × 1.10 = $302.50
75. [$105 − ($70 + $9)]/($70 + $9) = 33%
76. [($280K − $215K)/$215K]/3 = 10%
77. [$22 − ($130 + $9)]/($130 + $9) = −84%
78. 98 × $12 × 0.50 = $588
79. 18 × $12 = $216. The band got the guaranteed $300.
80. $123 × $(1.02)^3$ = $130.53
81. $69.99 + $17.98 = $87.97
82. $80 − $15 = $65
83. $100 − $48 + $25 − $30 + $48 − $1.42 = $93.58
84. 81 × $0.10 + 29 × 0.25 = $15.35
85. $1 × 3 × 3 × 3 × 3 × 3 = $243
86. Mya gave her mother $\frac{1}{3}$ of $15, which is $5. She gave her brother 40% of $15, which is $6. She gave her sister $4. Her brother got the most. Her sister got the least.
87. $768.44 + $79.40 + −$69.67 = $778.17
88. $986.60 + $62.59 + 82.68 = $1,131.87
89. $171.50 + $26.79 + −$86.29 = $112.00
90. $857.11 + −$89.23 + −$29.92 = $737.96
91. 11 gal × ($1.69/gal) = $18.59
92. $15/($2.25/wk) = 6.67 wk. 7 wk × ($2.25/wk) = $15.75. It will take 7 weeks, and she'll have $0.75 left over.
93. $10,000,000 − $3,800,000 = $6,200,000
94. Worse off. 0 − $55 = −$55
95. (99 × $0.99) − (99 bottles × $4.50/bottle) = −$347.49
96. $300 × (SFr6.35/$1) = SFr1,905
97. 4 × 4 × $0.25 = $4.00
98. A second 50-cent piece, 1 quarter, 2 dimes, 3 nickels, 5 pennies. There may be other correct answers.
99. £18.50 × ($1.81/£1) = $33.49
100. ¥5,413 × ($0.0034/¥1) = $18.40

Daily Warm-Ups: Math Word Problems

Answer Key

101. ¥73,850 × (€1/¥134.690) = €548.30
102. Three $10 bills, two $5 bills, one $1 bill. There may be other correct answers.
103. Two more 50-cent pieces, three dimes, two nickels, six pennies. There may be other correct answers.
104. $1000 − $100 = $900. $900/5 = $180
105. Six quarters, two dimes, five nickels. There may be other correct answers.
106. Three $100 bills, five $50 bills, two $10 bills, six $5 bills, three $1 bills. There may be other correct answers.
107. ($24.80 + $4.00)/4 = $28.80/4 = $7.20
108. Two $10 bills, two $5 bills, two $1 bills, three quarters, one dime, three nickels, three pennies. There may be other correct answers.
109. $1 + $1 × 3 + $1 × 3 × 3 + $1 × 3 × 3 × 3 + $1 × 3 × 3 × 3 × 3 + $1 × 3 × 3 × 3 × 3 × 3 = $364
110. $200/$1,500 = $\frac{2}{15}$ = about 13.3%

Counting

111. 120 − (91 − 75) − (88 − 75) − 75 = 16
112. g = number of girls. g + (g − 33) = 895. 2g − 33 = 895. 2g = 928. g = 464. There are 464 girls.
113. There are 4 ways. 1 + 1 + 11, 1 + 3 + 9, 1 + 5 + 7, 3 + 5 + 5
114. f = father's age. a = Agnes's age. f = 6a. f + 4 = 4(a + 4) = 4a + 16. Substituting, 6a + 4 = 4a + 16. 2a = 12. a = 6. f = 36. Agnes's father is 36. Agnes is 6.
115. 2 by 18, 3 by 12, 4 by 9, 6 by 6
116. $1 + 2^2 + 3^2 + 4^2 + 5^2 + 6^2 + 7^2$ = 140 oranges
117. (14 − 2)/(30 − 2 + 2) = $\frac{2}{5}$
118. 534/(534 + 507 + 88) = 47%
119. (0.25 × 28 + 4)/28 = 39%
120. ($\frac{1}{3}$ × 36 − 1)/(36 − 4) = $\frac{11}{32}$
121. 69/301 = 0.229
122. 1 + 2 + 3 + 4 + 5 + 6 + 7 + 8 + 9 + 10 + 11 = 66
123. ($\frac{1}{5}$ × 50 + 5)/(50 + 5) = $\frac{3}{11}$
124. two layers of cans; each layer, three by four cans
125. 28 − 9 + 14 = 33 rabbits
126. 15 oz × ($\frac{1}{3}$) = 5 oz
127. 537 ants × 6 legs/ant = 3,222 legs

Daily Warm-Ups: Math Word Problems

Answer Key

128. 30 dozen/case × 12 eggs/dozen × 20 cases = 7,200 eggs
129. 656 + 98 − 56 − 253 = 445 cards
130. 438/24 = 18.25. He needs 19 packages to store them all.
131. 2 by 40, 4 by 20, 5 by 16, 8 by 10
132. 2 by 45, 3 by 30, 5 by 18, 6 by 15, 9 by 10
133. 3 rows × 18 holes/row × 5 seeds/hole = 270 seeds
134. 18 or 2 by 3 by 3
135. $24 - (\frac{1}{2}) \times 24 - (\frac{1}{6}) \times 24 = 8$ retrievers
136. 530 turkeys/25 turkeys per container = 21 with a remainder of 5. Twenty-two containers are needed to transport all the turkeys.
137. 36 − 10 − 9 − (−1) = 18 tomatoes
138. 24 − 3 − 5 − 10 = 6 worms
139. 5 rows × 20 holes/row × 3 seeds/hole = 300 seeds
140. 6 rows × 10 holes/row × 3 seeds/hole = 180 seeds
141. 2 × 12 = 24 cookies on hand.
    (3 cookies/guest) × (25 guests) − 24 cookies on hand = 51 cookies needed
142. $\frac{27}{30}$ = 0.90 = 90%
143. $\frac{34}{40}$ = 0.85 = 85%
144. $\frac{40}{53}$ = 0.75 = 75%
145. 175 pix/(4 pix/page) = 44 pages. She needs two albums (64 pages) and there will be 64 − 44 = 20 pages left.
146. 3 rolls × 36 pix/roll × 0.07 = 7.56, or about 7 or 8 pictures
147. 3 + 7 + 7 − 2 = 15 heads of cabbage
148. $7{,}000 - (\frac{2}{3}) \times 7{,}000 = 7{,}000 - 4{,}666 = 2{,}334$. Approximately 2,300 letter carriers.
149. Form a 9-by-7 or 7-by-9 rectangle.
150. (984 − 984/2)/3 = 164 cards
151. 3.5 in × 2.5 in × 438 = 3,832.5 in^2 = about 26 ft 7 in square
152. 438 × 3.5 in = 1,533 in = 127 ft 9 in
153. 1 by 2 by 40, 1 by 4 by 20, 1 by 5 by 16, 1 by 8 by 10, 2 by 2 by 20, 2 by 4 by 10, 2 by 5 by 8
154. 40, or 2 by 2 by 10, or 2 by 4 by 5
155. 125, or 5 by 5 by 5
156. 27, or 3 by 3 by 3

Daily Warm-Ups: Math Word Problems

Answer Key

157. $2,000 + 12 × $844 = $12,128
158. 1 card = 3 in × 5 in = 15 in^2. Rectangle = 2 ft × 5 ft × 144 in^2/ft^2 = 1,440 in^2. 1,440 in^2/15 in^2 = 96 cards
159. 44,800 bushels/320 acre = 140 bushels/acre
160. Least: 1 + 8 = 9. Greatest: 24 + 31 = 55
161. 1 + 2 + 3 + 4 + 5 + 6 + 7 + 8 = 36 stickers
162. 30 papers × 6 papers/wk × 52 wk/year = 9,360 papers/year
163. 81 × 4 – 5 = 319
164. There are six possible formations: one row of 12, two rows of 6, three rows of 4, four rows of 3, six rows of 2, twelve rows of 1.
165. 1 + 2 + 3 + 4 + 5 + 6 + 7 + 8 + 9 + 10 = 55
166. Arabella reached a mile on the 14th day.

Day	Yards
1	500
2	600
3	700
4	800
5	900
6	1,000
7	1,100
8	1,200
9	1,300
10	1,400
11	1,500
12	1,600
13	1,700
14	1,800

167. 2 + 3 + 5 = 10. 2 × 3 × 5 = 30
168. Aiko is a doctor and is 35. Basil is a minister and is 40. Camilla is a lawyer and is 45.
169. 20 × 18 + 1 + 2 + 3 + 4 + 5 + 6 + 7 + 8 + 9 + 10 + 11 + 12 + 13 + 14 + 15 + 16 + 17 = 513 seats
170. 5 × 80 – (88 + 85 + 60 + 95) = 72

Daily Warm-Ups: Math Word Problems

Answer Key

171. Moira is 14 this year.
172. Gaspar is 35 this year.

Probability

173. 20 greens/75 gumballs = $\frac{4}{15}$, or about 27%
174. (20 greens/75 gumballs) + (19 greens/74 gumballs) = $\frac{581}{1,110}$, or about 52%
175. (20 greens/75 gumballs) × (19 greens/74 gumballs) = $\frac{38}{555}$, or about 7%
176. (75 gumballs − 20 greens)/75 gumballs = $\frac{11}{15}$, or about 73%
177. (30 reds + 20 greens)/75 gumballs = $\frac{2}{3}$, or about 67%
178. 13 hearts/52 cards = $\frac{13}{52}$ = $\frac{1}{4}$ = 25%
179. 4 kings/52 cards = $\frac{4}{52}$ = $\frac{1}{13}$ = about 7.7%
180. The probability that the first card you pick will be a king is $\frac{4}{52}$. If you get a king on your first pick, the probability of getting a king on your second pick is $\frac{3}{51}$. The probability of picking a king both times is $\frac{4}{52} \times (\frac{3}{51})$ = $\frac{12}{2,652}$ = $\frac{1}{221}$, or about 0.45%.

Daily Warm-Ups: Math Word Problems

Turn downtime into learning time!

Other books in the Daily Warm-Ups series:

- Algebra
- Algebra II
- Analogies
- Biology
- Character Education
- Chemistry
- Common English Idioms
- Commonly Confused Words
- Critical Thinking
- Earth Science
- Geography
- Geometry
- Journal Writing
- Mythology
- Physics
- Poetry
- Pre-Algebra
- Prefixes, Suffixes, & Roots
- Shakespeare
- Spelling & Grammar
- Test-Prep Words
- U.S. History
- Vocabulary
- World Cultures
- World History
- World Religions
- Writing